LLC

A Complete Beginner's Guide To Limited Liability Companies

(Get Your Business Up And Running In No Time)

Stuart Broad

© 2017

Executive Summary

The age-old selecting entities for starting a business has always narrowed down to a threefold one: sole proprietorship, partnership, and the corporation. Even though some states have oddities such as the Massachusetts Business Trust (the entity was recently done away with), the central entities have always remained sole proprietorship, partnership and the corporation in virtually all the countries for some time now.

But today, not only do several countries allow you to switch your partnership into a "Limited Liability Partnership (LLP)," there is rather a new form of business that has emerged on the scene and is gaining acceptance. That business entity is the Limited Liability Company (LLC). The era of LLC is no doubt a new phenomenon that that has already caught up with us.

Today, LLCs are being formed by almost every professional—accountants, lawyers, insurance agents; the list is endless. But what, you may be wondering, is the Limited Liability entity? Is it the same as a corporation? Definitely, it isn't a corporation. Is it a partnership? Somewhat, yes since LLC has certain characteristics that mirror the partnership entities. Or, is it a sole proprietorship entity? Absolutely no, even though it is sometimes regarded as one for tax computation purposes.

The chief purpose of coming up with this book is to help you get a complete big picture view of LLCs. Specifically, you'll understand the pitfalls that have been left by the uninformed entity creators. These pitfalls are many and dangerous to the legal health of the LLC entity. By understanding how the LLC is formed and managed, you'll be in a better position to invest wisely.

Let's get started.

CONTENTS

Chapter 1: Overview of LLCs

Limited Liability Companies (LLCs) has become a powerful tool for realizing several asset protection goals. The LLC is the most multi-faceted and flexible strategy for owning any property such as rental property which can help you insulating your business from dangerous assets as you promote your bottom line.

This section is designed to provide practitioners with a basic overview of LLCs which is a relatively new form of business entity in the US that provides owners the combined benefits of the pass-through taxation and limited liability. At the end of the chapter, you'll have a big picture view of LLCs including the brief history of LLCs in the US, the comparison between LLCs and other forms of businesses and merits and demerits of operating LLCs.

Before we start off, let us answer the fundamental question: "What is an LLC?"

What is an LLC?

A limited liability company is a structural business system where the members of the company can't be held personally accountable or liable for the company's debts and liabilities. LLCs are essentially amalgam entities that combine the properties of corporations and partnerships and some elements of sole proprietorships. In fact, the Pennsylvania Department of State defines an LLC as some sort of hybrid structure that lies between partnerships and corporations.

LLCs are relatively new forms of business structures that were first authorized by the state of Pennsylvania in 1995. Their business structure gives the liability protection of the company with the merits of being treated as a partnership. In particular, the LLC offers limited liability to its members—who are the owners of the company—and can be formed by more than one member with no restrictions on the maximum number.

While the characteristics of LLC makes them more like partnerships and corporations with provisions of management flexibility and the benefits of pass-through taxation, the merits and demerits to organizing the company as an LLC can vary depending on the type of business.

More specifically, the LLC grants protection from liabilities of corporations without the formalisms of the corporate minutes, bylaws, shareholders and directors. In contrast to the corporate law that allows the shareholders and other officers to be individually sued if the corporate structural formalities aren't followed, the LLC legislation doesn't take this route.

Compared to partnerships which are mostly used by professional organizations, LLCs can be formed by any group of people as contrary to partnerships that are only run by professional entities. For these professionals, the partnerships provide only somewhat limited protection from the liability since a professional who is a partner in a partnership remains personally liable for his/her own malpractices and gross negligence.

A brief background of LLCs

The first state in the US to formally recognize LLCs was Wyoming that allowed them in 1977. The Florida state followed suit in 1982. It wasn't until the 1988 IRS—

Internal Revenue Service—ruling that LLCs began to be treated as separate entities during tax computations and many businesses started to utilize them. The single activity triggered a wave of legislative activities throughout the US where each state sought to provide its businesses the opportunity to incorporate this new form of business entity.

By 1996, all the US 50 states and even the District of Columbia had enacted laws recognizing LLCs and developing frameworks that cover their formations, operations, and dissolutions. Even though LLCs are formally accepted by the IRS for tax computation purposes, LLCs are purely creatures of the state law. An LLC is an entity created by adhering to the procedures and regulations in the state of operation.

In the majority of states, forming the LLC requires the preparation of just one document, the so-called "Articles of Organization" which must be filed by the Secretary of the State and other designated state agencies. The Articles of Organization, which is usually brief sets forth the general background information about the LLC such as the name, the address, agent, terms and whether it will be run by members or managers hired by the members.

Every US state has a comprehensive set of rules for what should be contained in the document and also provides a fill-in-the blank form that must meet the statutory requirements. These days, the LLCs are no longer new forms of business but tested legal entities. They have now been recognized in all the US 50 states with well-established case laws and statutes.

The primary purpose of the LLC legislation is to enable individuals to create a legal entity that forestalls several of the tax and firm problems that are inherent in the sole proprietorships, corporates and partnership structures. The intent of the legislation is to allow persons who are members of LLCs to conduct their financial and business operations efficiently and conveniently devoid of formalisms and liabilities accruing from other business entities.

In the state of California and some other states, the LLCs must also file their document with the state of operating agreements—which are similar to partnership agreements—to provide a blueprint for the running of the organization the financial obligations of each LLC member and how the profits and losses will be shared. As a matter of fact, virtually every LLC will want to have a well-drafted Article of Organization operating agreement whether it isn't required or is required.

To the extent that an operating agreement doesn't exist or is silent on a particular operational issue, the courts have been resolving disputes by referring to each state's default legislative provisions that govern the operation of LLCs. An LLC can be classified to compute Federal income tax as either a partnership or even a corporation.

A domestic LLC that has at least 2 members will automatically be classified as a Partnership to compute Federal income tax unless the LLC chooses to be treated as a corporation. This can be done by filing an appropriate IRS Form. However, unlike partnerships, none of the LLC members are personally liable for the company's debts. Practically, all LLCs operates just like limited partnerships without the requirement for general partners.

Unlike an S corporations, the LLC has no restrictions on the number of shareholders, the classes of their stock, or even type of shareholders. And because in most cases the losses pass through to the LLC members, the LLC can be an attractive option for corporate investors and wealthy individuals. But, note that not all LLC will be eligible to pass-through taxation. This determination is usually made by applying the rules that have been outlined in an appropriate section of IRS. Even if the LLC doesn't qualify for tax treatment as a partnership or corporate entity, there are advantages for selecting this business form.

How does LLC compare with other business entities?

The most typical forms of business entities are the sole proprietorships, partnerships (which can include the limited partnerships), corporations, and the S corporations. This section delves deeper to provide you with an overview of the major features of these different business entities and compare them to LLCs. Let's dive in.

#1: Sole Proprietorships

A sole proprietorship business is an unincorporated business entity that is owned by one person. At present, it is the straightforward form of business organization to begin and maintain. The business has no lawful existence that is separate and apart from the individual owner. Its liabilities will be the owner's personal liabilities, and the owner must undertake the risks of the business for all of the assets owned whether they have been used in the business or not. The owner must include the income and expenses of the firm on his/her own tax return.

There are also no special forms to file with the state agencies to form a sole proprietorship.

Depending on the site and the type of business, however, the sole proprietor can be required to get a business license or even file a fictional business name statement. The sole proprietor can also sell all or some part of his/her business at any moment without agreement from others. As the business expands, the sole proprietor can also opt to change the form of the business, either through incorporation, filing to become an LLC or perhaps even adding a new partner.

The sole proprietor—being the sole owner of the business—has the complete flexibility to alter the nature of his/her business at any moment so long as it conforms to the statutory legal requirements of the new form of business entity.

#2: Partnerships

A partnership is any business that is operated by two or more individuals. The partnership can be a legally distinct entity from its members, or it can be a mere amalgam of its partners depending on its type. Other business entities like the corporation or even the LLC may also enter into the partnerships for the sole purpose of engaging in a new and related business.

There are two main categories of partnerships
- General partnerships
- Limited partnerships

Let us explore these forms of partnerships.

a) *General Partnerships*

In a general partnership, all the partners are "general members." This implies that each partner has unlimited liability for the business debts and can suffer obligations on the business's behalf within the scale of the business. Each partner acts as an agent for all the other partners of the business and thus, has the power to bind the organization through his/her actions including execution of contracts.

General Partnerships have few legal requirements. A general partnership doesn't require any written agreement. It can be formed with nothing more than just a general understanding or even a handshake between the partners. Each partner must intend to form a business relationship with the other partners the partnership to be created.

Just like the LLCs, without any written agreement, the state partnership laws must be drafted to govern partnerships. Each US state—with the exception of the Louisiana—has its own statutory laws governing the operations of partnerships. These legislations are usually called "The Uniform Partnership Act, " and in some cases, they are referred to as: "The Revised Uniform Partnership Act." These statutory laws set forth all the basic rules that apply to partnerships and are used to control many aspects of partnerships.

Partnerships aren't taxable entities. Each partner must include his or her share of the business's income or loss on the tax return. The partnership may allocate profit and losses agreeably, not just on the stock ownership that happens in S corporations. The allocation can also change from time to time. The partnership can accept and distribute their property without being subjected to tax computations.

The IRS code allows partnerships to convert to a corporation without involving taxation measures if the incorporation has been done properly. However, unlike the corporations, partnerships generally dissolves when one partner dies or withdraws from the business. Partnership agreements can, however, give alternatives for liquidation after the dissolution process like buyouts of a deceased or withdrawn member, election of the new general manager and continuation of the business by the partners who have remained.

Just like other forms of business entities, Partnerships can also be converted into LLCs. The conversion of the partnership business into LLC which has been classified as a partnership to compute federal taxes doesn't terminate the partnership. It is vital to note that the conversion process isn't a sale, exchange, or liquidation of any of the partnership interests and the partnership's tax year doesn't close.

In fact, the LLC can remain to use the partnership's taxpayer identification number.

Obviously, one of the main demerits of partnerships over LLCs is that all the partners will be liable for the debts and obligations of the business. This isn't true with the LLPs.

b) *Limited Liability Partnerships (LLPs)*

Limited Liability Partnerships have one or more general partners that have similar liability and authority as the case with general partnerships and one or the more "limited partners." The LLPs are no more than the amount of capital that a partner invest in the company. Generally, LLP partners can't take part in running of the partnership, or they become subject to general partners' liability.

Because of this feature, LLPs may make an LLC to be a more attractive option that allows all the owners to participate in the management of limited liability. However, in many US states, the LLPs provide only limited protection, shielding only against the liability that arises from malpractice committed by other partners in the firm.

While each state has both general and LLP acts, partners can decide to create their own rules in the partnership agreements. When these rules are written by partners, they override most of the stated' default partnership statutory provisions. In the absence of any statutory provisions, profits and losses are always divided equally among the partners. Death and withdrawal of any limited partner don't generally lead to the liquidation of the LLP.

The deceased partners' LLP interest may be passed on to their heirs or successors. However, unlike the general partnerships, LLP partners must file a certificate and a written agreement with the secretary of state or relevant state agencies.

c) Corporations

Corporations are legally distinct entities owned by their shareholders. The shareholders elect the corporation's board of directors and don't actively participate in the day-to-day running of the corporation. The board of directors makes the majority of corporate decisions where the corporations' officers that are appointed by the board of directors perform the corporation's day-to-day management.

A single individual can be the corporation's owner and the sole director and serve as any officer as required by law. In forming the corporation, the prospective shareholders can transfer money, property, or even both for the corporation's capital stock. In such instances, the corporation will take the same deductions just like sole proprietorship business to calculate its taxable income. The corporation may also take special deductions in such scenarios.

The total profits of the corporation can be taxed to both the corporation and the shareholders if profit has been distributed as dividends. However, the shareholders can't deduct any loss for the corporation. Most of the large firms operate as corporations which are the most accustomed entity and are closely regulated by legislations.

Corporations are usually taxed as separate legal entities unless they are S corporations. Under the current federal income tax law—that applies to the majority of states— corporations are usually taxed on their net incomes which are defined as the gross income less the allowable deductions. The corporate tax rates range from 15% to 35%. The non-cash properties can also be taxed as part of the corporation unless the group of people contributing that are contributing the property owners at least 80% of the corporation.

If the corporation dispenses money or other properties like dividends to their shareholders, then the shareholders are taxed on the distributions.

d) S Corporation

Eligible Domestic Corporations can evade double taxation—the first taxation to the shareholders and the second taxation to the corporation—by choosing to be treated as an S corporation. An S corporation is exempt from the federal income tax in US states. The shareholders incorporate on their tax returns the value of their shares as separately stated items of the incomes, losses, deductions, and credits.

The shareholders generally opt to form an S corporation when the corporation is profitable and dispenses nearly all of their profits to the shareholders and when the corporation incur substantial losses where the shareholders want to deduct from their personal income tax returns. In these circumstances, the profits and losses must be assigned based on the share ownership for tax computation purposes. The shareholders of the S corporation should include on their personal tax returns the profits of the S corporation even when no money was dispensed to them.

To qualify for S corporation status, the corporation must fulfill the following conditions:

- It should have not more than 100 shareholders that are individuals, certain tax-exempt firms, certain trusts or estates and for which none of them are nonresident aliens.
- It can issue only one class of its stock.
- It can't own 80% or more of another corporation.

Because all the shareholders of the S corporation should be individuals the S corporation
Can't raise funds from the venture capital funds, the corporations or even institutional investors. The one-

class-of-stock requirement inhibits the firm from issuing cheap founders' stock for the key employees. Most of the corporations that raise their money from outside sources must issue two classes of stock: the convertible preferred stock for the investors and the common stock for their employees.

The common stock is usually issued at a small fraction of the cost of the preferred stock since it lacks the liquidation processes, the dividends, voting regulations and other preferences of the preferred stock. Therefore, the S corporations are commonly used for family or other closely related owned business entities that obtain their capital from individual shareholders or debts from the outside sources.

Important consideration when selecting a business entity

Every business venture—whether LLC, LLP, sole proprietorship— is unique. If you are the owner or an investor, then no doubt you'll have different goals and concerns when selecting a business the type of business venture to invest in. As a matter of fact, no single business entity is perfect for any type of business.

However, there are some factors that can guide you when choosing the appropriate structure for your business entity.

The S corporations, partnerships, and LLCs all offer flow-through incomes for their owners (which my opinion, is desirable especially when these owners haven't been employed) when they don't want to reinvest their significant profits in the organization.

Therefore, one of the first sticking issue when determining whether the LLC is a suitable business form is whether the shareholders expect the business to generate profit or not.

If the firm expects to lose money initially, the flow-through entities such as the partnerships, LLCs or S corporations are preferable since it will allow the owners to deduct the losses from the taxable income. Secondly, the level of participation the investors want to have in the management of the company can affect the decision of selecting the suitable business entity.

The different forms may also restrict the level of involvement if the owners wish to maintain the limited liability status. With the LLC, the members can choose who will manage their entity. For instance, all or none of the members can choose to engage in the management process. Many other factors can also come into play when choosing the best business entity for each particular business entity. These include:

- Termination. How easy will it be to close down the business and return the capital and profits to owners? A sole proprietorship, for instance, can quickly be sold without consent from others.

- Cost: What is the minimum cost of setting up the business entity? The simpler the business structure, the smaller the cost.

- Ease of operation: Are there many formalisms linked with the business structure? The corporations, for instance, will demand a significant amount of formalisms, including the meetings, record-keeping and in some instances, public disclosures.

Advantages and disadvantages of LLCs

Here are some advantages of LLCs:

a) ***Tax Flexibility***. The IRS doesn't consider the LLC to be a separate entity for the tax computation purposes. This implies that the IRS can't tax the LLC directly. Rather, the LLC members get to determine how much they want to be taxed. There are many options:

- Single member LLC. This structure is taxed just like the sole proprietorship. The profits or losses from the business entity will not be taxed directly but rather will be taxed using the single personal federal tax returns.

- Partners in the LLC. The members choose to be treated as a traditional partnership for the tax computation purposes.

- LLC filing as a corporation. The members of the organization can also opt to file as if they were corporations.

b) ***There is less paperwork***. Compared to the S-Corps, the LLCs are very flexible. You'll want to have the LLC operating agreement so that you create rules that govern your business entity. Otherwise, your firm will be administered by the default laws in your state. With less strict requirements for compliance and necessary paperwork, the LLCs are easier to form and easy to keep in perfect legal standing.

c) ***Limited Liability***. As with corporations, the LLCs give their members protection from liabilities. This means that LLC members will not be personally liable for debts and other court judgments that may be incurred by the LLC. The creditors are also foreclosed from looking for the personal assets of the LLCs members. This is a meaningful shield that is not provided in the sole proprietorship or traditional partnerships.

Here are disadvantages of LLCs:

a) **Self-employment taxes**. Except you select to be taxed like a large corporation, the LLCs are subject to self-employment taxes. This implies that the profits of the LLC can't be taxed at a corporate level, but must go through to members who account for profits on their individual federal tax returns. These taxes are usually higher than they would be at the corporate level. The personal members can pay for federal items such as Medicare and Social Security.

b) **Confusion about roles**. While the corporations have particular roles such as directors and managers, the LLCs generally don't. This makes it difficult for the LLC to know who is in charge and who can sign certain contracts.

c) **_Limited life_**. In many jurisdictions, when a member leaves the LLC, it ceases to exist. This is unlike the corporation whose identity isn't affected by the comings and goings of the shareholders.

Chapter 2: Establishing an LLC

You've just completed your business strategy, and now the real fun begins where you start putting your concepts into motion in order to start a company. You already know that you should form a legal entity to help protect your assets. However, you are probably not sure the kind of business entity to form. You've possibly heard that the Limited Liability Company (LLC) is the easiest and fastest company to set up, but you're not sure if the LLC is the right choice for your business.

This section is designed to provide practitioners with a basic overview of LLCs that provides owners the combined benefits of the pass-through taxation and limited liability. At the end of the chapter, you'll have a big picture view of LLCs including the process of forming the LLCs. By the end of the chapter, you'll have a big and complete picture view of all stuff that is necessary for forming an LLC. Are you ready?

Formation of an LLC

The LLC, while accepted by the IRS for tax calculation purposes is still a product of the state laws. Therefore, in determining the mechanisms required to form LLC, you must examine the statutory laws of the state in which you want to operate. All the 50 states and the even the District of Columbia recognize LLCs and have statutory regulations that govern the formation and operations of LLCs.

As an illustration, let us examine the formation requirements in the state of California and compare it with requirements of other states where appropriate. Obviously, the first step in forming an LLC—which applies to all the states—is filing the Articles of the Organization with the Secretary of State or relevant state agencies. I know you're now asking, "What is an article of Organization?"

What is an Article of Organization?

An article of Organization is a document which is similar to the articles of incorporation that outlines the initial statements that form a limited liability company

(LLC) in many of the US states. In some states, the Article of Organization is referred to as a certificate of organization.

Your LLC's Article of Organization will act as a charter to form the existence of your LLC in your state and set forth basic information about the new entity you are creating. Stored as a single business document with the Secretary of State office or some state agencies, the Articles of Organization describe the basic identification and operational characteristics of LLCs. Once filed and approved, the Articles of Organization becomes a legally binding document that signifies the registration of your company as an LLC.

So, what is usually incorporated in the Articles of Organization?

No matter the type or even size of your new business entity, most of the states demand that the LLC's Articles of the organization must include, at a bare minimum, the following information:

- The new LLC's unique name and the address (the name and address must be principal places of the business).

- The nature of the LLC's business activities (these are usually stated in a broad language such as "to engage in a lawful activity," in order to evade restricting the LLC's business prospects).

- The name and address of the LLC's registered agent. The agent must be authorized to physically accept the delivery of certain legal documents (including the lawsuits) on behalf of the LLC.

- Name(s) of the manager(s) and all members of the LLC, if they are known at the time of filing.

- The identity the organizer(s) of the LLC that can initiate the organizational processes and be responsible for signing the articles of the organization before filing them with the state. If the articles names the manager(s) of the LLC, the manager(s) should also be required to sign the articles of the organization prior to being approved.

Whether you prepare the Articles of Organization yourself or consult experienced attorneys for assistance, the new LLC's Articles of Organization don't necessarily have to be complicated and extensive. If you choose to write them yourself, then your Articles of Organization will be most likely accepted for filing in the state so long as they have the minimum information that is identified the bullet-points.

However, you don't need to create the articles from scratch, as most of the states' Secretary of State Web sites have pre-printed Articles of Organization forms that you can easily pick and fill in a short notice. Once the new LLC's Articles of Organization is written and signed, it is ready for filing with the Secretary of State office. In all the states, the filing of the Articles of Organization will demand payment of a corresponding filing fee which varies depending upon the state of the organization.

Naming the Limited Liability Company

There are a couple of issues associated with the process of naming of a limited liability company. The first issues relate to the restrictions on the types of the names that are allowed and other requirements for what should be included in the name. The next primary issue is the process of reserving the desirable name in the company's state of the operation. Let's examine these issues.

Name Requirements and Limitations

All the states require the LLC to identify a unique name as such. For instance, the California Corporation Code Section 17052 demands the name of each LLC to contain either the words "Limited Liability Company" or use the abbreviation "LLC" or even "L.L.C" as the last words of the company name. All the states also allow for the acronyms of certain words in the name of the business entity such as "limited" and the word "company" can be abbreviated to "Ltd." and "Co." respectively.

Similarly, California and other US states forbid the use of entity names that are too similar to those of the existing businesses or those that would otherwise confuse the customers.

Other specific conditions under the California statutory laws include:

- The LLC can adopt a name which is similar to but isn't the same as the name of the existing domestic or even foreign LLCs if the company is existing LLC consents to that in writing to the Secretary of State in writing. The consent letter must be submitted on the letterhead of the concurring LLC and signed by an authorized officer of that LLC.

- The LLC name shouldn't falsely or imply any governmental affiliation.

- All the names should only use English alphabet or Arabic numerals (0, 1, 2, 3, 4, 5, 6, 7, 8, and 9) or a combination of Arabic numerals and English alphabet. The Roman numerals will be treated as letters and not translated into their equivalent numeric. The symbols are not allowed in any entity name; except the use of ampersand "&" character that can be used in the entity name as a conjunction.

- The name of LLC can't include the words "bank," "trustee," "trust," "incorporated," "Inc." in the California Corporations Code Section.

- The name of LLC shouldn't include the words "insurer" or even "insurance company" and any words that suggest the business is about issuing insurance policies.

The majority of states including California, explicitly allow the name of one or more LLC members so that it can be used in the name of the business entity. Additional naming rules can apply in some states which allow the professional LLCs. For instance, in Idaho, the name of a professional LLC must terminate with the words "Professional Company" or the acronym "PLLC" or "P.L.L.C." The District of Columbia demands that such entities have the words "professional limited liability company" and use the abbreviation "PLLC" or "P.L.L.C."

Reserving the company name

Many US states allow the organizers to reserve the unique name of the LLC by filing of Articles of Organization. A few other states such as Georgia, require the organizers to select their unique name before the filing of the articles. Some states also have the special forms and other regulations that are applicable to professional LLCs. Recall that reserving the unique name of LLC with the proper state agency such as the case of California and other states where the Secretary of State isn't the same concept as filing fictitious business names.

Organizers should contact the city and county clerk where the principal place of business entity is located for data regarding the process of filing and registering fictitious business names. The first step in naming the

LLC is to check the availability of the desirable name. In most states, including the California, a registry of all the existing names is maintained to be referred to by organizers. In the state of California, inquiries can be made in writing to the Secretary of State. Some other states have existing names already published online to help speed up the registration processes.

Once you find out that your desired name hasn't been used or reserved and can't cause confusion in the marketplace, the next phase is to reserve the name. Each state has a particular restriction on how long the LLC name can be reserved before filing the Articles of Organization. In California, the LLC name can be reserved for a maximum of up to 60 days. Also, each state generally charges a small fee for the right to reserve the unique name.

Fees and Reporting Requirements

All the states demand organizers to pay a small fee for registering the LLC. The fee is usually charged for the filing away of the Articles of Organization and differs greatly from state-to-state as set forth the statutory regulations. Some states such as Pennsylvania, charge higher fees for registration of professional LLCs. States

STUART BROAD

also levy a registration fee to foreign Limited Liability Companies who want to register in their states. This fee is generally, but not always, similar to the initial filing fee.

Some states also levy an annual fee. For instance, New York State bases its annual fee on the maximum number of LLC members. At the time of writing this book, the New York's annual fee was $100 per LLC member with the minimum fee being $500 and the maximum set to $25000.As with fees, the reporting requirements differ greatly from state to state. The majority of states don't require annual and biennial reports to be filed with the US states.

Some states demand annual reports while others require the biennial reports. Each state which requires the report sets forth the details of what must be included and when it should be filed. For instance, under Georgia statutory laws an annual report should be filed with the Secretary of State between 1st January and 1st April of each year. The report should include the following:

- The name of the LLC.

- The street address and county of the registered LLC office and the name of the registered agent in the state.
- The mailing address of the principal location of business
- Any additional data as may be required by the Secretary of State which is necessary for registration purposes.

Any foreign LLC that wants to do business in a given state is also subject to the filing requirements. A foreign LLC is one which is organized in another country or another state. These entities must file their reports if they want to engage in business within the different and foreign state. Outside any annual or biennial reporting which may be required, every state also sets forth in the statutes the types of records that can be maintained by LLC. All the members have a right to access these records pursuant to statutory provisions of the state law.

Chapter 3: Members' Responsibilities, Rights, and Duties in LLC

The flexibility in the ownership of LLC makes them a desirable business structure for several reasons. LLCs have more options in when it comes to structuring the membership responsibilities, rights, duties, and interests as other business entities. This chapter delves deeper to provide you with a big picture view of members' responsibilities, rights, and duties in LLC. Let's jump in.

Number of members

All the states allow single-member LLCs, making them a realistic business model for the sole proprietorship businesses. Also, there is no upper limit on the maximum number of the members that the LLC can have (contrast this with the S Corporations, for instance, that have limited number of 100 shareholders).

Types of membership

There is generally no limitation on who can hold the membership interest in an LLC. LLC incorporates the following type of membership:

- Individual
- Corporation
- Partnership

The only exception to this rule is the certain professional limited liability companies that may restrict the type of membership in the LLCs. With the S corporations, on the other hand, have significant limitations on the types of shareholders that are allowed; while corporations and partnership and other specialized groups are forbidden from owning the shares in an S corporation.

Similarly, an LLC is generally free to have various types of classes of ownership and membership interests such as differences in the voting rights. Such rights are usually spelled out explicitly in the LLC's operating agreement. Each state has particular provisions of statutory law that governs the creation of different classes of membership. For instance, in California, the Corporations Code Section (17102) stipulates that the

Articles of Organization should provide for the creation of membership classes having relative rights, powers, and even duties as the Articles of Organization including the rights, authority, and duties of members.

Membership interests

Member's interest in the LLC consists of both economic and non-economic interests. The economic interests are majorly two-fold. First, the member's capital contribution to the LLC that the member can withdraw under certain conditions. Second, the member can enjoy the right to receive the profits at the end of a financial year. The member's interest can also grant him/her the abilities to contribute to the management of the company.

Members' interest is usually transferable to other persons or entities, even though the management rights of the transferees can be limited. Each LLC's operating agreement procedures can have provisions that alter the basic statutory principles.

For instance, a particular LLC can provide that a certain class of member doesn't have the right to contribute to the management, or can vest the management authority in a single member. In these circumstances, the operating agreement may also provide that the membership can't be transferred without the unanimous consent of other members.

Rights of the members

As with the shareholders of a corporation, LLC members have certain rights, including the rights to notice of the meetings of the organization and the rights to inspect the business entity's books and all other financial records such as tax returns. The membership rights are crucial in situations where the members hold a minority interest in the LLC and are at odds with those that are holding the majority ownership.

Each of the state's statutory scheme must set forth the specific rights of the members, which must include the right to liquidate their investments. The members of LLC have all the statutory rights to access data about the companies they own such as the rights to inspect the books and other relevant records. Ideally, under the Uniform Limited Liability Company Acts in most states (ULLA), members have the following rights:

a) LLCs must provide members and their agent's access to their records, if any, at the company's office or any other reasonable locations that have been specified in the operating agreement. The LLC must also furnish former members and their agents' access to proper records concerning the period of which they were members. The right of the access offers the chance to inspect and copy the records during ordinary business hours. The company can impose a reasonable levy based on the costs of the labor and materials being requested.

b) LLC must furnish a member and any other legal representative of the deceased member or members under the legal disability:

- Without request data regarding the company's business and state of affairs required for the proper exercise of that member's rights and the performance of the member's functions as contained in the operating agreement.

- On request, other data pertaining the company's business state of affairs, except to the extent that the demand for the data demanded is unreasonable or otherwise inappropriate under the circumstances.

c) Any member has exclusive rights upon written request to be given to the LLC and obtain any of the company's written operating agreement.

One of the vital areas of the legislation of LLCs that has yet to be operationalized concerns the rights of minority members. This issue may arise in many forms such as the situations where the minority member or members have been being "rammed out" of the company's decision-making by other members that have a controlling interest in the business entity. The law of corporations—as it is currently constituted in the majority of states can be used to resolve the outstanding disputes arising from the minority owners of LLCs in the absence of defined laws controlling the business entity. In some states, minority members can seek judicial dissolution of LLC in case certain issues have arisen that are bring conflicts in the organization.

Duties of members

Under the common law which is provided for in statutory regulations in many jurisdictions, the directors and officials of the corporation are bound by duty to the corporation and all the other shareholders and always act in the best interests of the LLC.

Also, in the process of executing their duties, the directors and officials are similarly required to make well-informed decisions backed by data about the business. Both of these legal principles of the corporation are also applied to LLCs.

As with the corporate law that is often used to resolve the disputes in LLCs, its basic principles relating to the director's duty of the loyalty and duty of care will be commonly carried over in figuring out whether the member of LLC has breached his duties to other members or even the company as a whole. Therefore, the general import of these principles of corporate law is crucial in understanding the responsibilities of LLC members specifically the members managing the company.

Recall, however, that one of the cornerstones of LLCs is its flexibility. Thus, an LLC'S operating agreement must specify the explicit rules governing the duties and responsibilities of their members in detail. Each state also provides particular rules in statutory regulations that dictate the minimum standards of conduct from members. In some instances, these standards can vary between the member-managed and the non-member-managed companies.

However, as long as the standards have been set forth in the operating agreement don't violate the state statute, the courts will always uphold them in the event that a litigated dispute arises. For instance, the Florida law is representative of this type of statutory regulations adopted by many states concerning the conduct of members of LLCs:

- Each manager and all the managing members must owe the duty of loyalty and the duty of care to the LLC and all of the members of the LLC.
- Refrain from dealing with the LLC in the conduct and winding up of the LLC business as or on behalf of the party that has an interest in the LLC.
- Refrain from competing with the LLC in the conduct of the LLC's business before the dissolution of the company.

LLC membership as a security

Membership in an LLC is regarded the personal property of the LLC member. Contrast this to the corporation where the "owners" hold the "shares" in the corporation that are treated under the state and federal law as a form of security. Now, whether the federal securities laws apply to the membership in an LLC, relies on the nature of each business entity. In particular, the more control that members have over the day-to-day running operations of the business, the less likely that their membership can be classified as security.

The most important fact that comes to mind when classifying membership interests in an LLC as security for the purposes of the federal law is the act of triggering of registration and other disclosure requirements that are involved in the purchase and sale of that security. Each state can also have a statutory regulation that provides the nature of membership interests in the LLC. Recall, however, that the federal law preempts this form of security. This implies that even if the interest isn't treated as security for the purposes of the state law, it can very well be under the federal law.

Some states expressly specify the membership in an LLC as security for purposes of the state law. Some of these states include Michigan, Alaska, New Mexico, Nevada, Ohio, and Vermont. The Maine state, on the other hand, explicitly excludes memberships in the LLC as security for the purposes of state law. The Texas state expressly exempts only the professional LLCs. Other states have a statutory regulation frameworks for determining the conditions under which the membership can be treated as security.

Transfer, sale, and assignment of memberships

Because LLCs are governed by the state laws, look first to the legislations of the state of the firm to determine the rules concerning the transfer, sale or even assignment of the membership in an LLC. Second, identify the business entity's own articles of organization to check what, if any, restrictions the members have developed. Aside from any restrictions contained in the applicable state law, the members are usually free to figure out their own rules for the process of transferring, selling and resigning via their operating agreement.

A number of states have enacted laws that demand the unanimous consent of all the members before the membership can be transferred. Some of these states are Alaska, Alabama, Delaware, Arizona, Colorado, Florida, and Georgia. Some other states, including the state of California and Connecticut, demand that the majority of the members of the LLC must consent to the assignment of the interest contained therein. To the extent that any ownership interest in the LLC is considered as security, additional regulations can apply to transfers.

Dissociation of members

Each state has its provisions that govern when and under what conditions a member can resign, or disassociate from the LLC (note that the term "disassociate" is often used whenever any member decides to leave the LLC, whether it is on voluntarily or involuntarily basis). These can also be changed by the business entity's operating agreement. Some states give a notice period before the member can withdraw unless otherwise provided in the statutes. For instance, the Alabama State requires at least 30 days written notice before a member resigns. A few other states such as Maryland demands six months' notice.

Members may also resign by giving not less than six months' prior written a notice to other members of LLC at their respective addresses as indicated on the books of the LLC unless the operating agreements don't specify whether the member has the right to withdraw. Also, if the operating agreement dictates another time or other circumstances of withdrawal, members' resignation request can be turned down.

Some states such as Alaska, explicitly prohibit members from resigning prior to dissolution process unless provided for in the business entity's operating agreement. A few other states such as California and Arizona allow members to resign at any time they wish. For instance, upon written notice, the Kentucky state forbids a member from withdrawing without the consent of all LLC members unless the operating agreement provides specifies.

The North Dakota statutory law, on the other hand, dictates that LLC members have the power, though not the right to resign, implying they can do so but only subject to the damages. The Nebraska and Rhode Island states totally fail to address the issue of member's withdrawals, leaving it totally to the discretion of the business entity to so provide terms in the operating agreement.

Recall that the ability of the member to resign from the LLC can be a vital consideration when selecting a business entity. For some individuals, the liquidity of a firm can be an important issue when choosing to create and invest in a business entity. Obviously, to the extent that the Limited Liability Company is member-managed, the successful continuity of the firm will always be frustrated without limitations on the resignation.

Chapter 4: Management of Limited Liability Companies

This chapter delves deeper into the world of LLCs to provide you with a bigger picture view of managerial issues in LLCs. By the end of this chapter, you'll be in a position to understand the key managerial issues that you have to deal with if you opt to start an LLC. Are you ready?

Designating the management of an LLC

One of the several merits of LLCs over other business entities is the flexibility that it provides in deciding who can manage the enterprise. For instance, in an LLP, the limited partners are always precluded from participating in the day-to-day running of the organization if they want to continue maintaining their status as Limited Liability partners. An LLC, however, can be managed by some, all or even none of the members of the LLC, and yet in all of those circumstances, all the members maintain their status as limited liability company members.

There are two basic types of LLCs when it comes to management:

- Those LLCs that are manager-managed
- Those LLCs that are member-managed.

The articles of organization usually specify the business entity's management structure; and indeed, some states demand the articles give that provision. Most of the states presume that in the absence of the statutory provisions, all the members will work as managers of the organization.

Some states, such as the state of Tennessee provide that the LLC can be managed by the board of governors, akin to the concept of the board of directors in a corporation. For instance, the Tennessee Code of Regulations provides the following:

If the LLC is member-managed, then all the powers will be exercised by or under the power of, and the business affairs of the direction of its members.

If the LLC is board-managed, the all the authority will be exercised by the business and affairs of the LLC must be handled by the direction of the board of governors who are subject to the provisions statutory regulations. Unless otherwise specified in the articles or organization or the operating agreement, each governor must have the equal voting power.

Some LLCs can also designate titles such as that of the "officer" that is usually linked with the corporations. Because LLCs are free –subject to some constraints in the state law – to create their own management structures, they are equally free to create the job titles and vest the authority as they may deem appropriate.

In the member-managed LLCs, each member has the equal rights to the management and performance of the company's business unless it has otherwise been provided in the operating agreement. The LLC is free to assign the voting rights based on any agreeable formula it wishes, in the absence of a specific statutory law to the contrary. LLCs usually assign voting rights based on the per capita basis –where every LLC member has one vote—or based on the capital contribution—where the stakeholders are assigned votes that are proportional to their ownership interests in the LLC.

For most of the decisions, the majority vote of the LLC members is enough. Certain decisions, though, pursuant to either the state law or the LLC's operating agreement, must be unanimous to go through. In the manager-managed LLC, the members don't have rights to participate in the day-to-day running of the organization unless they are serving as manages which is subject the different rules of the LLC's operating agreement. In most instances, managers will not be required to be the members of the LLC, and the company is free to define its own criteria for managers in the organization.

Managers as agents of a limited liability company

The common law regulations of the agency law typically apply to the managers of an LLC.

This implies that the manager has the express authority to bind the company and create other legal obligations on behalf of the business entity such as executing the lease on a particular building or buying a piece of equipment that can be used by the business entity.

Most of the states have the particular statutory regulations that detail the power of a manager to act as an expressive agent of the LLC which take effect in the event that other

Legal provisions aren't defined in either the LLC's articles of organization or the articles of operation.

For instance, the Alabama state law provides—pursuant to its Title 10—as follows:

a) Except as provided in the regulations of the articles of the organization or operating agreements, every member of an LLC is an agent of the company for the sole purpose of its business or state of affairs, and any act of a member such as the running the organization in the name of LLC or any instrument carrying out the activities of the business entity.

b) If the articles of the organization specify that the management of the LLC is vested in the manager or a group of managers, both of the following conditions can take effect:

- No member that is acting solely in the capacity as the LLC member can act as an agent for LLC.

- Every manager of an LLC is an agent of the company for the sole purpose of its business and state of affairs, and the act of the manager, including, and not limited to, the running of the organization in the name of the LLC of any instrument for carrying out activities of the company that binds the LLC.

- An act of the manager or any member that is not for the purpose of carrying on in the general way the business doesn't bind the LLC unless it has been authorized in the operating agreement at the time of the transaction or at any other given time.

- No acts of the manager or any member in contravention of a limitation on authority will bind the LLC to people having full knowledge of the limitation.

- In the LLC that is managed by its members under the statutory provisions or articles of organization, the only fiduciary duties that an LLC member owes to the company and to other LLC members will be the duty of loyalty and the duty of care.

A basic synopsis of the common statutory law principles that have been set forth in the American Law Institute's Restatement of Agency (ALIRA) is helpful in figuring out the statutes and the contract provisions that affect this area of management. There are 3 types of authority that are recognized in common law:

- Express
- Implied
- Apparent

Let's dive in and explore these types of authority.

#1: Express Authority

The Express Authority is the simplest and authority that is prevalent in LLCs. Principal, whether in an LLC or corporation specifically gives the agent the expressive authority to engage in certain decisions on behalf of the organization. And since the power or the authority is spelled out clearly, both for the agent and the principal will clearly define their roles with regard to the day-to-day running of the Limited Liability Companies.

#2: Implied Authority

The Implied Authority is that power which is implicit with regard to the position of the agent. For instance, it can simply be implied that the manager of a company has the sole authority to hire and fire his/her subordinates. The fact that the authority hasn't been expressly granted by the owners of the LLC doesn't negate the existence of the power.

#3: Apparent Authority

The Apparent authority exists under certain conditions when a specific third party believes that the purported agent has expressive authority to bind the principal, whether he does it or not. The Apparent Authority is usually specified in the Restatement of the Law of Agency as the power to affect the legal relations of other people by transactions with the third-party persons, professedly as agents for the other that arises from and in accordance with the other people's manifestations for such third-party persons."

Under these conditions, the principal will have taken some actions that can reasonably cause the third-party person to believe that the person in question is, in fact, the agent of the organization. When third-party individual acts in accordance with the apparent authority of that agent, the principal will always be bound by the actions to the extent that he/she would if the agent had the expressive authority.

The Uniform Limited Liability Company Act (ULLCA)

According to the Uniform Limited Liability Company Act (ULLCA), which was adopted by the National Conference of Commissioners (NCC) on Uniform State Laws, the basic principles of the agency that governs the general partnerships can also apply to the LLCs.

Two basic principles have been outlined in ULLCA:

- Member-managed companies
- Manager-managed companies

Let's jump in and explore these principles.

#1: Member-managed Company

In the member-managed company, each LLC member automatically becomes the agent of the company and thus has the express authority to bind the LLC when acting in the course of the company's business operations. The only exception to this rule is when the LLC member has no authority to take the actions of running the organization in question. This fact must be known to the LLC member at the time he or she is acting on behalf of the company.

Unless specifically provided in the operating agreements by other members, the LLC member generally doesn't have the express power to bind the LLC in a matter that is not within the usual scope of its business.

#2: Manager-managed Company

Members generally don't have the express authority to bind the LLC unless they are also serving as managers.

The rules that govern managers will be the same as those that apply to the members of the member-managed companies which we have described above. The managers can bind the LLC when acting within the general scope of the business unless they lack the power and this fact is already known to the third-party. Also, unless specifically authorized by the LLC, the manager doesn't have the authority to bind the LLC in any action that isn't within the normal scope of the business entity. For instance, the manager of an LLC that has organized to operate a restaurant can't have an expressive authority to lease an airplane on behalf of the LLC, unless he or she has been specifically authorized to do so.

Chapter 5: Professional Limited Liability Companies

The many merits of LLC status such as pass-through taxation, limited liabilities, the flexibility in the management structures makes them a desirable option for several types of businesses. One area that is particular where the LLCs have become popular and advantageous is for the professionals, who historically used the partnerships as their preferred entity.

The Professional LLCs –or simply PLLCs—are allowed in virtually all the states and even the District of Columbia. Only the California state prohibits the PLLCs. A professional is generally any individual who is licensed by the state to perform a specialized service. A professional can be a doctor, a dentist, an accountant, an attorney, an architect or even a scientist.

The most apparent merit for the group of professionals to create an LLC rather than a partnership is the evasion of personal liabilities. This is particularly the case with regard to professionals such as the doctors and lawyers that face increased prospects of professional liability lawsuits such as malpractice. Although the members are always protected in general from the debts and any liabilities that accrue from the enterprise, they are not protected against the suits that may arise out of their own malpractices.

Also, because most states allow the single member LLC's, the sole practitioner can also have the benefit of the protections provided by a Limited Liability Company rather than the more complicated route of creating a professional corporation. The majority of states expressively authorize the professional limited liability companies. These states include Arizona, Alabama, Georgia, Idaho, Connecticut, Kentucky, Maine, Maryland, Arkansas, Florida, and Iowa. The Professional Limited Liability Companies are also expressively authorized in the District of Columbia.

The Professional limited liability companies are also implicitly allowed in the following states: Colorado, Alaska, Delaware, Illinois, Indiana, Hawaii, Louisiana, New Mexico, New Jersey, Oklahoma, South Dakota, Pennsylvania, South Carolina and Wisconsin. In spite of this widespread approval of the Professional LLCs, there are still some hurdles on the kind of businesses that can be formed of this type of entity in some of the states.

For instance, the state of Illinois forbids LLCs from engaging in banking activities. The states of Arizona, Delaware, Oklahoma, Kansas, and Pennsylvania also bar limited LLCs from engaging in both banking and insurance industries. The state of California– in addition to forbidding the PLLCs– also prohibits the formation of banking, insurance, and trusts PLLCs.

For most of the decisions, the majority vote of the PLLC members is enough. Certain decisions, though, pursuant to either the state law or the PLLC's operating agreement, must be unanimous to go through. In the manager-managed PLLC, the members don't have rights to participate in the day-to-day running of the organization unless they are serving as manages which is subject the different rules of the PLLC's operating agreement. In most instances, managers will not be required to be the members of the LLC, and the company is free to define its own criteria for managers in the organization.

Naming and forming a PLLC

Each state specifies the data that must be included in the LLC's articles of organization and operating agreements. Additional data is usually required for the PLLC, including the fact that the business entity is a limited liability company and the complete description of the specific professional service that is being rendered to the public. The states are explicitly authorizing professional LLCs demand the business entity to identify itself as a Professional Limited Liability Company in its name. For instance, the Texas state provides that a PLLC can adopt a unique name not contrary to the statutory provisions of the law or ethics governing the practice of these companies.

In particular, the state of Texas requires the name of the LLC to contain the words "Professional Limited Liability Company." The name can also be abbreviated as "P.L.L.C." or simply "PLLC' and should include other words that are prescribed be required by law.

Limitations on membership and practice of PLLCs

#1: The single professional rule

The majority of states restrict the PLLCs to the practice of a single profession. Therefore, the attorneys and accountants can't be members of the same PLLC even if they are allowed pursuant to statutory provisions. Such PLLCs will most likely violate the professional regulations of conduct that preclude the professionals from being partners with those who aren't licensed.

Nonetheless, in Texas, for instance, certain medical practitioners are allowed to be members of the same LLC even when they aren't technically in the same profession, such as podiatrists who can be members of the same PLLC with medical doctors. If you are a professional, you should recall that irrespective of what is provided in statutory regulations, you must always abide by applicable professional laws of conduct or you can face disciplinary proceedings by the licensing agency of their states.

#2: Membership Restrictions

Unlike the other types of LLCs where there are no limitations on who can become an LLC member, the PLLCs are always closed to the non-professionals as members. The majority of states bars people who are not members of the profession from being members, managers or officials of the company. Equally, a membership interest in the PLLC can only be transferred to another person who is licensed in the same profession. This also implies that only individuals can be members PLLC, contrary to other types of LLCs where, for instance, the corporation or the partnership can have membership in an LLC.

When a member becomes disbarred or loses their license to practice the specific profession, they should cease to be a member of the PLLC immediately. The majority of states generally demand the PLLC to repurchase the ownership interest from that former member under these situations. The PLLC's articles of organization or the operating environment can contain a formula for operationalizing the valuation of the membership interest in these situations like when an attorney or an accountant withdraw his license to practice or retires. In that case, he or she can no longer be a member of the PPLC.

If the person who has withdrawn his license or retired was the sole member of the PLLC, he or she is normally allowed to continue acting in the capacity of manager or even a member for the limited purpose of winding up the company. He or she can, however, not continue to render or even provide professional services during that time. If a person who is not a professional receives an interest in the PLLC from a member like an attorney member of a PLLC dies and leaves his shares to his wife, then the PLLC must immediately buy that membership pursuant to a formula that is provided in the PLLC's operating agreement.

Liability for members Professional Limited Liability Companies

The LLC is an attractive option for a variety of reasons, not the slightest of which is a limited liability that provides for its members. In the case of PLLCs, liability is always limited, but not to the degree that it would be for other types of businesses. While members of the PLLC can't have personal liability for the debts and obligations of the company such as malpractices, they are usually legally responsible for their own actions.

Also, LLC members will still be subject to all the rules of professional responsibilities that govern their professions, such as the case of doctors and the doctor-patient privilege.

The majority of states that expressively recognize the have specific statutory regulations that address this issue. When a member becomes disbarred or loses their license to practice the specific profession, they should cease to be a member of the PLLC immediately.

The majority of states generally demand the PLLC to repurchase the ownership interest from that former member under these situations. The PLLC's articles of organization or the operating environment can contain a formula for operationalizing the valuation of the membership interest in these situations like when an attorney or an accountant withdraw his license to practice or retires.

Foreign Professional LLCs

Just like other forms of LLCs, the Foreign Professional Limited Liability Companies—or simply FPLLCs— must also comply with the special state registration obligations if they want to engage in business beyond their state of formation.

The Texas state, for instance, provides that an FLLC may apply for a practicing certificate of authority to do professional services in that state by applying its statutory provisions.

In this respect, the Secretary of State may not be obliged to issue the certificate unless the name of that FLLC or the name it elects to use in that state meets the minimum requirements of the Act. An FPLLC can render professional services in the state of Texas only through a member, a manager, an officer or agent described in the company's operating agreements.

Also, the certificate can't be issued to the FLLC under its statutory regulations unless the application for that certificate has a statement that the authority in which the LLC is organized will permit the reciprocal admission of the LLC if it were formed in this state. Similarly, a PLLC is generally free to have various types of classes of ownership and membership interests such as differences in the voting rights.

Such rights are usually spelled out explicitly in the LLC's operating agreement. Each state has particular provisions of statutory law that governs the creation of different classes of membership. For instance, in California, the Corporations Code Section (17102) stipulates that the Articles of Organization should provide for the creation of membership classes having relative rights, powers, and even duties as the Articles of Organization including the rights, authority, and duties of members.

Registration requirements for Foreign Professional Limited Liability Companies

Every state has particular rules and regulations governing the registration of a Foreign Professional Limited Company. The majority of states generally provide a standardized registration format with most states requiring the following information:

- The name of the Foreign Professional Limited Liability Company or, if its name is unavailable for use in the state, then an alias name that satisfies the minimum requirements of that state.

- The name of the State or even country under whose law the Foreign Professional Limited Liability Company is organized.

- The street address of the Foreign Professional Limited Liability Company's principal office.

- The address of Foreign Professional Limited Liability Company's initial designated office in the state.

- The name and street address of its current agent for servicing of processes in that state.

- Whether the duration of the Foreign Professional Limited Liability Company is for a specified term and, and if so, the specified period.

- Whether the Foreign Professional Limited Liability Company is manager-managed, and, if so, the name and addresses of each current manager.

Some states demand annual reports while others require the biennial reports. Each state which requires the report sets forth the details of what must be included and when it should be filed. For instance, under Georgia statutory laws an annual report should be filed with the Secretary of State between 1st January and 1st April of each year. The report should include the following:

- The name of the FPLLC.
- The street address and county of the registered FPLLC office and the name of the registered agent in the state.
- The mailing address of the principal location of business
- Any additional data as may be required by the Secretary of State which is necessary for registration purposes.

Revocation of registration of a Foreign Professional Limited Liability Companies

Just like the domestic companies, the state can revoke the authority of a particular foreign entity to engage in business activities in that state under certain conditions. These conditions are always outlined in each of the state's law and generally include the following:

- If the Foreign Professional Limited Liability Company fails to pay the required fees, taxes or even penalties.

- If the Foreign Professional Limited Liability Company fails to file the required reports.

- If the Foreign Professional Limited Liability Company fails to maintain the agent for servicing of processes.

- Determination has been made the Foreign Professional Limited Liability Company has made a material misrepresentation in applying for jurisdiction to operate.

Conclusion

Limited Liability Companies (LLCs) have become a powerful tool for realizing several asset protection goals. The LLC is the most multi-faceted and flexible strategy for owning any property such as rental property which can help you insulating your business from dangerous assets as you promote your bottom line.

But as we have seen in this book, the process of starting a Limited Liability may not be a walk in the park if you don't understand the laws and statutory regulation in each state. The aim of this book has been to help you get a complete big picture view of LLCs. Specifically, the book has covered all the aspects of understanding the pitfalls that have been left by the uninformed entity creators.

By understanding how the LLC is formed and managed, you should now be in a better position to start a Limited Liability Company. We hope that you've enjoyed reading the book. Good luck.